Sponsored Technology Workshops 2022

Held at the 2022 AIChE Spring Meeting and 18th Global Congress on Process Safety

San Antonio, Texas, USA
10 – 14 April 2022

ISBN: 978-1-7138-6834-7

Printed from e-media with permission by:

Curran Associates, Inc.
57 Morehouse Lane
Red Hook, NY 12571

Some format issues inherent in the e-media version may also appear in this print version.

Printed with permission by Curran Associates, Inc. (2023)

AIChE
120 Wall Street, FL 23
New York, NY 10005-4020

Phone: (800) 242-4363
Fax: (203) 775-5177

www.aiche.org

Additional copies of this publication are available from:

Curran Associates, Inc.
57 Morehouse Lane
Red Hook, NY 12571 USA
Phone: 845-758-0400
Fax: 845-758-2633
Email: curran@proceedings.com
Web: www.proceedings.com

TABLE OF CONTENTS

641678 Sponsored Technology Workshop - Enhance Reliability of Refinery Operations with Electrolyte Based Process Simulation Solutions - OLI Systems

Abstract:

Electrolyte thermodynamics play a crucial role in the material reliability of a refinery asset. As refineries focus on making products on targets, maximizing profits and reducing feedstock costs, asset reliability may be negatively impacted as we push operational boundaries. Attend this workshop to understand how chemistry insights can help increase reliability and extend asset life by setting Integrity Operating Windows.

During this workshop, the presenter will take a focused look at the chemistry of contaminants in crude unit overhead systems, and will, in detail, explain the phenomena that causes chemical corrosion and damage in the units. We will investigate the reasoning behind neutralizing amine and wash water injection points and their separation.

This workshop will deliver actionable insights on why chemical companies suggest particular neutralizing amines for the overhead systems and how refiners can use the same tools to assess the performance of chemical additives. Finally, we will review a real-life case where an opportunity crude was evaluated as an excellent source to increase profitability. However, the elevated level of contaminants in the crude caused the corrosion rate on the overhead, as measured by boot water analysis and iron content, to increase dramatically.

In this workshop we will cover:

- Phase change corrosion: Ionic Dew Point (sometimes called Initial Dew Point), Water Dew Point and -Hydrocarbon Dew Point Calculations
- Neutralizing amine selection criteria
- Wash water rate optimization
- Processing of high-contaminant crudes and desalter upsets
- Developing Integrity Operating Windows

641745 Sponsored Technology Workshop - Reach Your Sustainability Targets with New Levels of Efficiency and Innovation - Aspen Technology

Abstract:

In this workshop, you will learn about AspenTech innovations and how AspenTech University will enable you to maximize business performance and meeting sustainability goals. You will learn how to:

- Reduce use of energy, water and feedstock through process optimization and prioritized capital allocation
- Develop new processes and sources for future energy needs and decarbonize existing applications
- Achieve better Design and Sustainability Outcomes with Concurrent Engineering

641072 Sponsored Technology Workshop - P&ID's and Information Management – How they can play together - Open Range Engineering

Abstract:

An unfortunate trend in today's workflow is that once the P&ID has been released for construction, the integrity and data starts to get lost. As a manager you could pay upward of $2,000 for an updated P&ID and still end up with a piece of paper that does not achieve your end goals. In this workshop we can help you understand how to create a value-added P&ID that achieves ease of use, which in turn will create ownership for the PSI and PSM data at your facility and improve your auditing success. Most facilities information retrieval is sporadic and force you down a winding rabbit hole of document accessibility, this path that is usually followed is very time-consuming and confusing. We can show you how to use a simple P&ID to solve your information gaps and create a successful information management retrieval superhighway by using modern technology. When information retrieval becomes simple, it allows your personnel to achieve ownership of the information management requirements. This ownership is achievable through subject mastery which is easily realized by holding your information in an organized manner. This workshop would benefit anyone that needs to access vital information for their facility and wants to have a simple solution for information management.

641675 Sponsored Technology Workshop - Streamline Your Chemical Management Processes - Benchmark ESG

Abstract:

Looking to simplify & streamline your chemical management program? Join our workshop to learn how digital solutions can help site and business leaders organize, document, and manage chemical inventories, ensure proper reviews as part of new chemical approval requests, facilitate new supplier/product introduction, and standardize the management of change procedures in alignment with Process Safety Management regulatory standards and industry best practices. We'll touch upon critical themes related to the Process Safety Management Standards in addition to workflows surrounding SDS's, Management of Change, Process Hazard Analysis, Product Stewardship, Dangerous Goods Management, and much more.

641676 Sponsored Technology Workshop - Thermoplastic Design Considerations; Choosing the Ideal Material for a CPI Piping System - IPEX USA

Abstract:

This session will introduce different types of thermoplastics and discuss their unique and differentiating qualities. It will expand on relevant physical properties which allow thermoplastics to outperform traditional piping materials. Discussions will cover important design considerations which should be considered when selecting a material for an application. The workshop will also address common questions about the use and benefit of thermoplastics in industrial settings.

641677 Sponsored Technology Workshop - Can Digital Twins Solve our Industry's Workforce Challenges? - Voovio Technologies

Abstract:

Two former plant managers from BASF and DuPont will look at how Digital Twins can solve the challenges of the workforce and close this gap. Successful Digital Twins must capture tribal (people) knowledge and include the use of (procedure) simulators with data on competency. The ability to educate and test operators on equipment specific procedures provides valuable data on how reliable those operators are. You can measure your people's availability, reliability and productivity. But what about the 'operators' that operate and maintain the equipment? There is a growing knowledge and skills gap in our industry: experts leaving our plants due to retirements or turnover.

We will look at how Digital Twins can solve the challenges of the workforce and close this gap. Successful Digital Twins must capture tribal (people) knowledge and include the use of (procedure) simulators with data on competency. The ability to educate and test operators on equipment specific procedures provides valuable data on how reliable those operators are. You can measure your people's availability, reliability and productivity.

Join us to learn how our industry can close the knowledge gap with the adoption of Digital Twins.

641682 Sponsored Technology Workshop - Dynamic Barrier Management: Integrating PSM into Operational Risk Management - Wolters Kluwer | Enablon

Abstract:

In the workshop it will be demonstrated how Process Safety Management can be seamlessly integrated with Operational Risk Management.

It will show how operational processes like Permit to Work, MOC, Job Hazard Analysis, Incident Management and Process Safety Management can be brought together to support day-to-day decision making taking into account the actual cumulative risk status of the facility.

Based on the 360 Risk concept the actual status of the facility can be viewed from various angles, from an asset integrity / engineering point of view, from an operational point of view and from a scenario (BowTie) point of view.

Allowing the involved departments to assess emerging issues and drilling down to the underlying source data to address them.

641694 Sponsored Technology Workshop - Fast Track Computational Chemistry with Turn-Key HPC in the Cloud - Rescale

Abstract:

Computational simulation continues to play an increasingly critical role in the understanding and development of products and processes within the life sciences and energy sectors.

Huge strides and advancements in high performance computing (HPC) in the cloud have unlocked acceleration opportunities for computational fluid dynamics, computational chemistry and molecular dynamic simulations to discover new drugs, materials.

Rescale provides access to over 100+ HPC architectures in a multi cloud environment to run both open source and commercial computational chemistry, molecular dynamics and computational fluid dynamic (CFD) codes. These include Nvidia GPU Architectures to further accelerate computational studies and reduce discovery and development cycles.

In this workshop, we'll provide a demo walkthrough of running computational chemistry codes such as GROMACS, LAMMPS on the Rescale Platform.

"Through Rescale, we were able to draw results more than 2x faster than our existing workflow, which helped us make better business decisions."
- Jerry Maeng, Managing Director, AZothBio

"Bionano's use of Azure on Rescale reduces compute cost for genetic disease analysis by more than 50%, compute time by 30%."
- Mark Oldakowski, COO, Bionano

641746 Sponsored Technology Workshop - Renewable Energy: Integrating Process Safety, Asset Integrity, and Risk Management - Siemens Energy Process & Safety Consulting

Abstract:

The global shift from fossil fuels to renewable energy production is occurring at an accelerated pace. Global energy leaders are adapting to both rising market demands and regulatory changes. While renewable fuel processing facilities are built and existing petrochemical plants converted, the impacts of these changes could introduce health and safety, environmental, and reliability risks.

This interactive workshop provides an overview of key renewable technologies and uses case studies to illustrate how this new energy landscape affects process safety, asset integrity, and risk management.

Register for the workshop here.

AUTHOR INDEX